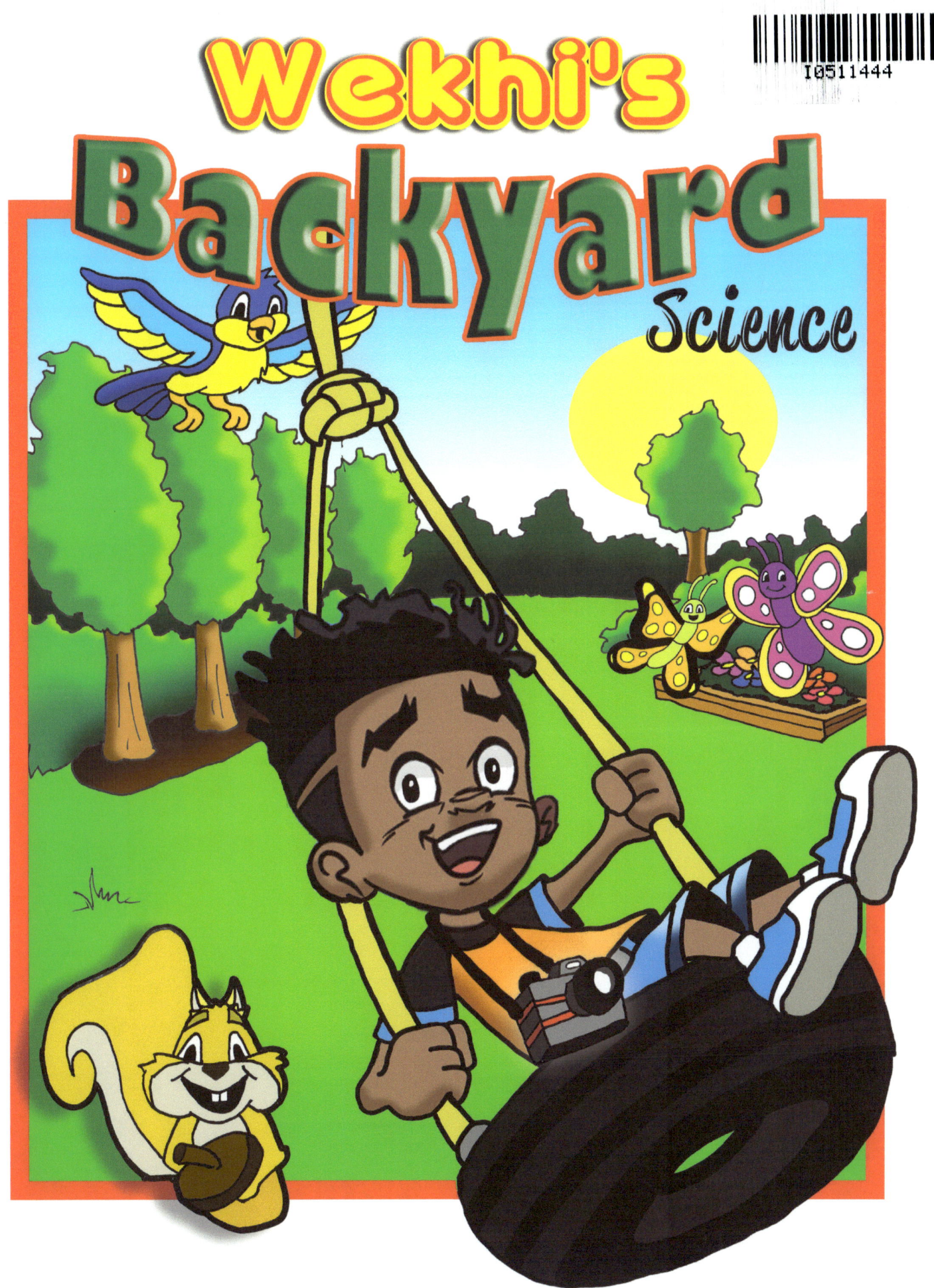

Copyright © 2022 by Delores Henriques

All rights reserved.

It was the beginning of summer, and Wekhi was very happy.
No school! He had big plans for his first week at home.
Seven days of fun!

His backyard is his most favorite place in the whole world.

Wekhi loves the tall, swaying trees.

He loves the birds that come to eat the berries from the trees and the bird feeders.

He loves the butterflies that follow the blooms on the dandelions and the hibiscus.

Wekhi loves the animals that come to nibble at the grass and weeds.

He loves the big oak tree. Its full branches are homes for many animals that live in his backyard.

Most of all, Wekhi loves his **giant, red treehouse**.

He would spend his time in his **giant, red treehouse** watching the animals that come and go.

Suddenly, he had a great idea! He would make this summer very special! The digital camera Grandma gave him for Christmas would come in handy. He could use it to snap pictures of all the animals he sees.

He would make a scrapbook of all his pictures.

On the first- day, Wekhi got up early. He got dressed, ate breakfast, got his camera, and went to his backyard.

What did Wekhi see?

A hungry caterpillar was eating a leaf, a spider was swinging from his web in a tree, and a small brown rabbit popped up from the grass and hopped away,

and his giant, red treehouse.

On the second day, Wekhi got up as soon as the sun peeked through his bedroom curtains. He got dressed, ate breakfast, got his camera, and went to his backyard.

What did Wekhi see?

He saw a woodpecker digging for food in the trunk of a tree and two green frogs jumping across the grass to catch the bugs

and his giant- red treehouse.

On the third day, Wekhi got up as soon as he heard birds chirping by his window. Then, he got dressed, had breakfast, got his camera, and went to his backyard.

What did Wekhi see?

The sunflowers' bright yellow flowers were in full bloom.
But then, along came the honeybees ... zoom, zoom, zoom

and his giant- red treehouse.

On the fourth day, Wekhi got up as soon as he heard the dog barking in the backyard. He got dressed, had breakfast, got his camera, and went to his backyard.

What did Wekhi see?

His dog Buddy was barking and digging for a bone he had buried earlier, frightened grasshoppers,

and his giant- red treehouse.

On the fifth day, Wekhi got up as soon as he heard the sound of squirrels. Then, he got dressed, had breakfast, got his camera, and went to his backyard.

What did Wekhi see?

He saw two squirrels playing in a tree. One scampered down to the ground, picked up an acorn, and ran back up again

and his giant- red treehouse.

On the sixth day, Wekhi got up as soon as he heard the chirping of birds. Then, he got dressed, ate his breakfast, got his camera, and went to his backyard.

What did Wekhi see?
Two orioles fed their babies wiggly worms in a nest in the tall willow tree.
A deer came by for a quick snack

and his giant- red treehouse.

On the seventh day, Wekhi got up as soon as he heard the parakeet's sounds flying by. He got dressed, ate his breakfast, got his camera, and went to his backyard.

What did Wekhi see?
Hummingbirds with beautiful wings kissed the wildflowers and fluttered their wings.
A snake slithered slowly by and,

his giant, red treehouse, the safest place to be.

The eighth day was special for Wekhi. He could sleep in, play with his toys, or read his comics or a book, but not today. Instead, he got all the pictures he needed and put the images in the scrapbook. Soon, he had all his photos in the scrapbook, created a cover,

and even wrote a title.

Can you guess the title?

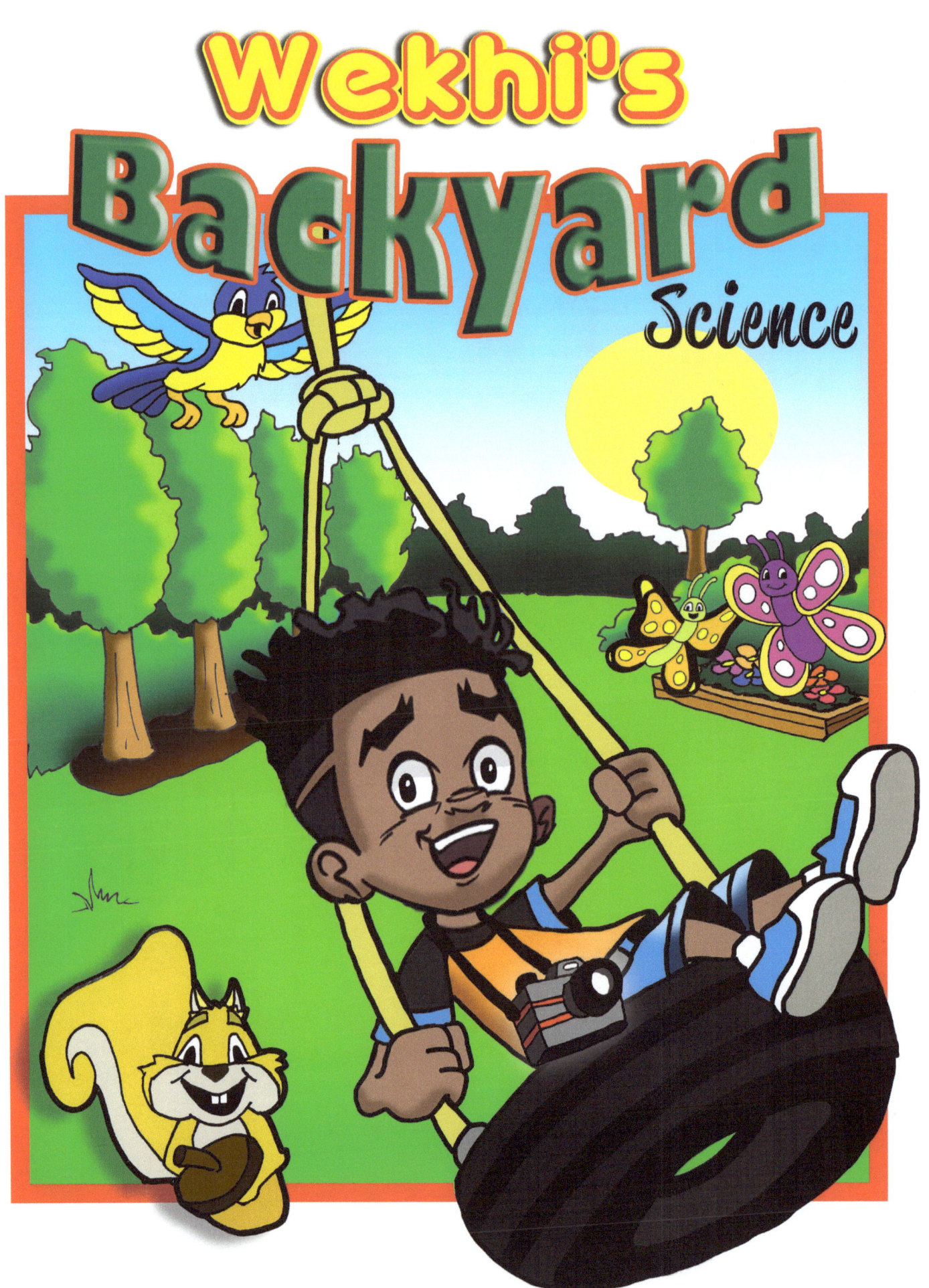